Spanish Explorers

Researching American History

introduced and edited by
Pat Perrin

Sixteenth century woodcut showing two oversized navigators sighting the sun with an astrobe (right), and a star with a cross-staff (left).

Discovery Enterprises, Ltd.
Carlisle, Massachusetts

First Edition © Discovery Enterprises, Ltd., Carlisle, MA 2003

ISBN: 1-57960-088-3

Library of Congress Catalog Card Number 2002116656

10 9 8 7 6 5 4 3 2 1

Printed in the United States of America

Subject Reference Guide:

Title: *Spanish Explorers*
Series: *Researching American History*
introduced and edited by Pat Perrin

Spanish Explorers

Credits:

Cover art: Detail from a mural by Newton Alonzo Wells depicting the history of the Mississippi River. This shows De Soto with his fellow travelers and Native Americans.

Illustrations: The illustrations on pp. 11, 16, 18, 33, and 54, 16th century after Le Moyne, engraver, De Bry, courtesy of Professor Troy Johnson, California State University, Long Beach. Found at www.csulb.edu/

Maps on pp. 7, 12 from Richard Hofstadter, William Miller and Daniel Aaron, *The United States: The History of a Republic,* Englewood Cliffs, NJ: Prentice-Hall, 1962, pp. 8, 12.

All other illustrations credited where they appear in the text.

Contents

About the Series

Researching American History is a series of books which introduces various topics and periods in our nation's history through the study of primary source documents.

Reading the Historical Documents

On the following pages you'll find words written by people during or soon after the time of the events. This is firsthand information about what life was like back then. Illustrations are also created to record history. These historical documents are called **primary source materials**.

At first, some things written in earlier times may seem difficult to understand. Language changes over the years, and the objects and activities described might be unfamiliar. Also, spellings were sometimes different. Below is a model which describes how we help with these challenges.

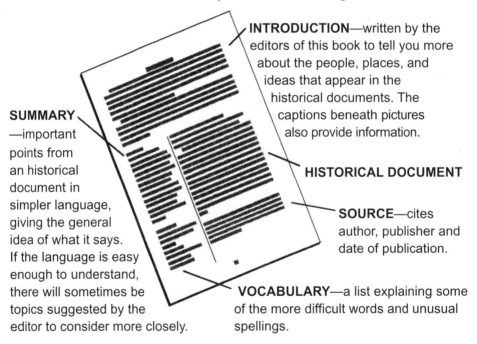

INTRODUCTION—written by the editors of this book to tell you more about the people, places, and ideas that appear in the historical documents. The captions beneath pictures also provide information.

SUMMARY—important points from an historical document in simpler language, giving the general idea of what it says. If the language is easy enough to understand, there will sometimes be topics suggested by the editor to consider more closely.

HISTORICAL DOCUMENT

SOURCE—cites author, publisher and date of publication.

VOCABULARY—a list explaining some of the more difficult words and unusual spellings.

In these historical documents, you may see three periods (…) called an ellipsis. It means that the editor has left out some words or sentences. You may see some words in brackets, such as [and]. These are words the editor has added to make the meaning clearer. When you use a document in a paper you're writing, you should include any ellipses and brackets it contains, just as you see them here. Be sure to give complete information about the author, title, and publisher of anything that was written by someone other than you.

Introduction: Reasons to Explore

Throughout human history, individuals and small groups of people have gone exploring, traveling by land and by sea. Sometimes they were looking for brand new territory, and sometimes they found it by accident.

Norsemen landed on the American continent as early as the 10th century. Stories say that Chinese and others might have visited even earlier. But in the 1400s, an explosion of European exploration began. The period is often called the Age of Discovery or the Age of Exploration.

Italian and Portuguese merchants and sailors took the lead in exploring new land and sea routes to Asia. But it was the Spanish who launched the first major investigations of the Americas—the lands that Europeans called the New World.

Of course, a spirit of adventure drew many explorers to undertake these dangerous voyages into little-known territory. But Europeans also had three other strong reasons to look for new lands.

- **Primogeniture** According to the rule of primogeniture, only the oldest son in a family could inherit much of his father's property. Younger sons—even from the nobility—had to find other means of support. (Women generally had to marry someone who could support them.) Some younger sons had the connections to get a sea voyage financed. Successful expeditions brought them fame, career advancement, and sometimes wealth.

- **Mercantilism** Countries needed sources of wealth, too. If a nation wasn't self-sufficient, it would soon be dominated by its neighbors. Trade made the difference. European countries could get rich by selling goods in Asia, Africa, and the Middle East, and by bringing back luxury items to sell in Europe. But trade became very competitive. The first explorers were looking for easier, faster routes to rich trading partners.

- **Religion** The desire to convert foreigners to Catholicism drove some to the New World. Most Spanish voyages were accompanied by at least one priest or friar (a member of a Catholic religious order).

In this book you'll find excerpts from the letters and journals of early Spanish explorers. These have been translated from the Spanish of several hundred years ago into modern English—so most aren't hard to read. Many sidebars contain commentary rather than a summary of the source material.

According to old tales, the oceans were inhabited by sea monsters and most other lands were either too hot or too cold for humans to live in. In this illustration, a ship is surrounded by flying fish.

Exploring the Known and the Unknown

During the Middle Ages, traders from Italian city-states such as Genoa and Venice found their way to Asia, Africa, and the Middle East. They brought back spices, chinaware, carpets, perfumes, medicines, and jewels. It was a profitable business. And traders also brought back exciting stories about the foreign lands they visited.

For example, in 1271, 17-year-old Marco Polo went to Asia with his father and uncle. Years later, he dictated the story of his travels. When the account was published, Marco Polo's descriptions of the court of the

An illustration of Marco Polo's visit to Cambay, India.

Chinese Emperor, Kubla Khan, and other wonders made many young men want to go exploring, too.

Venturing into the Oceans

Much of this early trading meant slow and difficult journeys across land. At first, sea voyages were kept short and close to shore. According to old tales, the oceans boiled in some places, and ship-eating sea monsters lived out there in the deeps.

A Portuguese prince called Henry the Navigator (1394-1460) refused to believe those scary old stories. Gathering all of the scientific information of his time, Prince Henry improved both ship design and navigation techniques. He sent ships along the African coast, where sailors discovered new places, peoples—and markets.

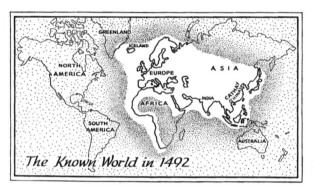

The Known World in 1492

By the 1490s, Europeans had explored the African coast and much of Asia. However, many people believed the Earth was much smaller than it actually is. Some thought they could reach the East quickly by sailing west. They didn't realize that two large continents were in the way.

Looking for a Shortcut

Merchants in Portugal, Spain, France, and England became eager to find their own sources of valuable trade goods. They especially wanted to find a faster route to the Indies. ("The Indies" meant India and all nearby areas.) The merchants hoped to bring luxuries back more cheaply than their competitors—or at least get in on the great profits to be made.

Many people thought that the Indies were just on the other side of the Azores (Portuguese islands in the Atlantic Ocean). If that turned out to be true, ships could get to the wealth faster by going across the Atlantic Ocean. Even though that meant traveling far out of sight of land, it could be worth the risk.

These early voyagers were going into unexplored territory. Although scholars believed that the earth was round, nobody had yet proved that. Nobody had actually sailed around the world. The journey turned out to be a lot longer than they expected. And the land they found wasn't the Indies at all, but a whole New World.

Crossing the Great Ocean

The first seagoing explorers stayed in familiar waters, often within sight of land. As early as the 12th century, sailors on the Mediterranean sea used simple magnetic compasses. But they'd need other tools out on the ocean.

Dead Reckoning

Early sailors (including Columbus) tracked their route by "dead reckoning." They judged their direction (by compass), speed, and distance from a known starting point. To judge speed, they threw a piece of debris overboard. The pilot used a memorized chant to count off the time it took the debris to pass between two marks on the ship's rail. Then the officer of the watch put a peg into a traverse board to mark their position. Later, they transferred their path to a map.

cross staff

Celestial Navigation

Sailors could determine their latitude (distance north or south of the equator) and the approximate time of day by measuring a heavenly body's height above the horizon. The Greek cross-staff was still in use when the Age of Discovery began. The navigator pointed a cross-staff halfway between the horizon and the sun or a star. Then he slid the crosspiece until one end was in line with the horizon, the other with the star. A scale on the staff gave the star's altitude.

A quadrant was a metal quarter-circle which had a weight on a string. The navigator would sight the North Star along one edge, and the hanging string would cross a point on a scale on the curved edge.

An astrolabe was a brass or bronze disk with a pointer fastened to the center. One person held the astrolabe while another sighted along it and set the pointer to show the angle of the sun or a star. The astrolabe was less accurate than a quadrant.

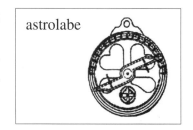

Spanish caravels (Courtesy of Mediasource)

Ships

Explorers crossed the ocean in wooden sailing ships. The smaller ones also had oars for rowing. Most carried cannons for defense against other ships, including guns designed to fend off native canoes.

Spanish caravels were trustworthy in the ocean and maneuverable enough to explore bays and rivers. Early caravels had triangular, "lanteen rigged" sails. Later caravels were "square rigged," with square sails, or with both types. Two of Columbus's ships—the *Niña* (49 feet long) and the *Pinta* (56 feet long)—were caravels. Some caravels were as long as 75 feet. Many of today's ocean-going liners are *wider* than those ships were *long*.

Larger, heavily armed galleons could cross the ocean with a lot of cargo. They could be more than 100 feet long. However, early galleons were top-heavy, and turned over easily during storms. The fleet of Hernando de Soto included seven galleons.

Pirates sometimes used faster caravels to catch Spanish galleons.

Smaller ships were used for exploring the shallower waters of bays and rivers. These boats also had oars and benches for men to row them. Some were flat-bottomed and could carry horses as well as soldiers.

Christopher Columbus

A sailor from the Italian city-state of Genoa, Christopher Columbus (1451-1506), wanted to open up new trade routes to Asia. Columbus had probably read Marco Polo's famous adventure stories, describing the wonders and wealth to be found in the East.

Like others of his time, Columbus came to believe that he could reach the Indies by sailing west. After years of trying to raise money for the journey, Columbus was finally financed by Spain. He called the islands he reached the "Indies" because he thought he had reached his goal. Now we call them the "West Indies."

Columbus made four voyages between 1492 and 1504. He began his journal of his first voyage with a review of his orders from King Ferdinand and Queen Isabella.

Consider this:

Columbus knew less about where he was going than the astronauts who walked on the moon. His journey was really into the unknown.

Commentary:

Columbus reached the Bahamas, but he was still half a world away from Asia. He made three more voyages, but never realized that an entire continent lay between the land he reached and the lands he had hoped to reach.

Vocabulary:

armament = the process of arming for war
determined = decided
disposition = nature
hitherto = before now
sufficient = as much as is needed; enough

Letter to Spain

Your Highnesses ... determined to send me, Christopher Columbus, to the ... countries of India, to see the said princes, people, and territories, and to learn their disposition and the proper method of converting them to our holy faith; and furthermore directed that I should not proceed by land to the East, as is customary, but by a Westerly route, in which direction we have hitherto no certain evidence that any one has gone.... Your Highnesses, in the same month of January, ordered me to proceed with a sufficient armament to the said regions of India.... I left the city of Granada, on Saturday, the twelfth day of May, 1492, and proceeded to Palos, a seaport, where I armed three vessels, very fit for such an enterprise, and having provided myself with abundance of stores and seamen, I set sail from the port, on Friday, the third of August, half an hour before sunrise ....

Source: *Medieval Sourcebook*, "Christopher Columbus: Extracts from Journal," at http://www.fordham.edu/halsall/source/columbus1.html

Magellan's ships off Mactan Island, in the Phillipines, where he was killed.

Other Early World Travelers

Meanwhile, other seagoing explorers were traveling the coast of Africa. Bartholomew Díaz sailed around the southern tip of Africa in 1486-87. Vasco da Gama reached India in 1498 and returned to Portugal the following year, his ship loaded with spices and jewels. And in 1519-21, Magellan made his remarkable world voyage.

An Italian named Americus Vespucius claimed to have seen the South American continent before Columbus got there in 1498. Vespucius' claim is in doubt, but he got a lot of publicity and in 1507 a geographer gave the new land the name "America."

The Spaniards

At the time of early voyages to the New World, the Catholic church was very powerful. In 1493, the Pope granted the discoveries that Columbus had made to Spain. The Spanish then had the privilege of exploring the new lands, and the responsibility of converting the native people to Christianity.

Spanish explorers established bases on Puerto Rico and Cuba. There, they developed a home base and set up shipbuilding operations.

In the early 1500's Spanish explorers investigated the New World. They established bases in the West Indies—from which captains led small bands of soldiers onto the mainland. Later groups included settlers, as well.

Lured by Legends

Spanish explorers were also drawn to the New World by legends of wealth. In South America, there was supposed to be a city of gold and jewels called El Dorado. When early explorers found the great cities of the Aztecs and Incas, the legends exploded. Some became convinced that the entire New World was rich with treasure for the taking.

Another legend told of the Seven Cities of Cíbola, grand and wealthy cities that were supposed to be in North America. And then there was the story of Prester John's kingdom, where rivers were filled with gold and a Fountain of Youth flowed magic waters. That one was said to be in Asia or Ethiopia, but some explorers came to believe that the fabulous fountain was in the New World.

Spanish conquistadores attacking natives. (Found at "Images of Iberoamerica," www.groups.msn.com/SpaininAmerica/indios)

Conquistadors

The Spanish soldiers were called conquistadors. They were well-armed, well-trained, and experienced in battle. Once they landed and left their ships, they usually had no way to turn back. So they fought ruthlessly to conquer any natives who challenged or resisted them.

The natives had never seen horses, guns, armor, or white men. Nevertheless, flint arrows could pierce chain mail and the skilled native archers of North America could inflict damage even on armored men. After their first battles, many conquistadors switched from chain mail to heavily padded cloth jackets, which worked better.

Ponce de Leon (Found at www.beaconic org/district/bayhistory/bhis6)

Ponce de Leon in Florida

The Spanish explorer Ponce de Leon was with Columbus on his second voyage to the New World. Ponce de Leon settled his family on Hispaniola (now the Dominican Republic) and retired there to run a plantation. But he came out of retirement when the Spanish crown selected him to colonize Puerto Rico and to explore the Bahamas.

At that time, people thought that Florida was a large island. In 1513, Ponce de Leon explored the Florida Keys and part of the west coast of the peninsula. He landed near what is now St. Augustine. In 1514, he received a royal commission to colonize the "isle of Florida," but he didn't get funding for the voyage until 1521. On that trip, he was wounded by an Indian arrow and had to return to Havana, where he died from his wound.

The Fountain of Youth

Natives in the Caribbean told explorers about a legendary fountain. They said that its waters would restore health and youth to whomever drank there. Some thought this must be the Fountain of Youth of ancient legend. According to most stories, Ponce de Leon was searching for this fountain in Florida. However, some historians believe that he was actually hoping for a spiritual rebirth and to gain greater honor.

De Leon's Letter to the King of Spain

Very Powerful Lord

As it has always been my usage and custom to serve the Royal Crown in these parts by command of the catholic King … among those services that I have mentioned I discovered the Island of Florida, at my own expenses and provisions, and others in its vicinity …. And now, I return to that island it be pleasing to the will of God. To make a settlement; being able to carry plenty of people with whom to do it …. And I intend to discover more of the coast of that island, and to know if it one…. I will send Your Majesty an account of that which may be done or seen in these parts where I may travel. I will ask favors, and from now I pray you will grant them, because I could not bear to undertake so great a voyage, nor of such expense. Neither could I be able to accomplish it except with the protection and favors from Your Caesarean Majesty…. From the Island of San Juan, the City of Puerto Rico which is in the Indies of the Ocean Sea. The Tenth Day of the Month of February of One Thousand Five Hundred and Twenty One Years.

Of your Majesty The Slave and Servant Who Kisses Your Very Royal Feet and Hands. Juan Ponce de Leon

Source: Florida of the Conquistador website at http://www.poncedeleonwater.com/letter

Summary:

As always, I serve the Crown. I discovered the island of Florida and others nearby at my own expense. I will return, God willing, and take people to settle there. And I plan to explore more of the coast and find out if it is an island. I will send you a complete report. And I ask your help because I cannot pay for such a voyage myself.

Consider this:

Why do you think Ponce de Leon referred to the king as "Your Caesarean Majesty"? (Where did the title of Caesar come from?)

Vocabulary:

usage = habit

The Adventures of Cabeza de Vaca

Alvar Nuñez Cabeza de Vaca (about 1490–1557) chose his unusual name—Cabeza de Vaca means "head of a cow"—from his mother's side of the family. The name came from a 13th-century battle between the Spanish and the Moors. A Spanish peasant found an unguarded way into the enemy stronghold. He marked the pass with the skull of a cow.

After the Spanish defeated their enemy, the grateful king created a new title for the helpful peasant—Cabeza de Vaca. The peasant's family became members of the nobility, and their fortunes improved.

As a young man, Alvar Nuñez Cabeza de Vaca joined the military. In 1527 he was made second in command of an expedition to Florida. They set sail with six hundred people in five ships. One was a smaller ship designed for exploring rivers and streams. The others were probably merchant ships called "naos" or "carracks." They were in between caravels and galleons in design and size.

Near Cuba, the ships were battered by terrible storms. The expedition finally landed in Florida with about 400 soldiers. The captain of the expedition, Pámfilo de Narváez, claimed the land for the Spanish empire.

Sixteenth century engraving of a Florida Indian village that might have been seen by the early Spanish explorers.

What happened over the following months and years was unplanned, unexpected, and unimaginable.

Landing in Florida

The Governor debarked with as many men as the ships' little boats could hold. We found the *buhios* deserted, the Indians having fled by canoe in the night. One of the *buhios* was big enough to accommodate more than 300 people; the others were smaller. Amid some fish nets we found a gold rattle.

Next day the Governor raised flags and took possession of the country in Your Majesty's name....

He then ordered the balance of the men to land, with the horses, of which only 42 had survived the storms and the long passage at sea; these few were too thin and run down to be of much use.

The Indians of the village returned next day and approached us. Because we had no interpreter, we could not make out what they said; but their many signs and threats left little doubt that they were bidding us to go. They, however, went away and interfered no further.

Source: Cabeza de Vaca, *Adventures in the Unknown Interior of America*. Translated and edited by Cyclone Covey. Albuquerque: University of New Mexico Press, 1998, p. 31.

Commentary:
Cabeza de Vaca called Captain Pámfilo de Narváez "the Governor."

Summary:
The Spaniards went ashore in their small boats. The Indians had left, but the Spaniards found a gold rattle. Narváez claimed the land for Spain. The 42 Spanish horses were in bad condition.

The next day the Indians returned and the two groups began communicating with sign language.

Vocabulary:
accommodate = hold comfortably
buhios = a type of wigwam with an open shed attached
comptroller = officer who handles financial matters
debarked = went ashore, disembarked

On Foot in Florida

Like most of the explorers, Captain Narváez was hard on the Indians. When they found some crude gold ornaments in one village, Narváez took the Indian leader hostage. But Chief Utica wouldn't tell the Spaniard where the gold came from. According to some accounts, the captain cut off the Indian's nose. Then the Indians made up a story about great wealth among a people called the Apalachee, who lived a short way to the north.

Florida natives paddling canoe by a wigwam.

Determined to find those riches, Narváez decided to split his people into two groups. One would proceed on land and the other by sea. He said they would join up later. Cabeza de Vaca objected, but was brushed aside.

Early in May, three hundred men marched into the interior of Florida. Forty were on horseback, but the horses were in bad shape from the sea voyage. The men would soon have worse problems—and they would never see their ships again.

Some of the Spaniards died from disease, probably malaria and dysentery. Others were killed by natives. (They quickly discovered that Indian arrows could pierce Spanish armor.) They found no wealth. Three months later, an exhausted, sick, lost, and somewhat smaller group of Spanish explorers found their way to the Gulf coast of northern Florida. But they had no way to leave.

No Way Out?

A third of our force had fallen seriously ill and was growing worse by the hour ... with death the one foreseeable way out .... Considering our experiences, our prospects, and various plans, we finally concluded to undertake the formidable project of constructing vessels to float away in.

This appeared impossible, since none of us knew how to build ships, and we had no tools, iron, forge, oakum, pitch, or rigging, or any of the indispensable items, or anybody to instruct us. Worse still, we had no food to sustain workers.

Source: Cabeza de Vaca, *Adventures in the Unknown Interior of America.* Translated and edited by Cyclone Covey. Albuquerque: University of New Mexico Press, 1998, p. 45.

The Building of the Barges

Next day one of our men [said] he could make wooden pipes and deerskin bellows.... We also instigated the making of nails, saws, axes, and other tools we needed out of the stirrups, spurs, crossbows, and other of our equipment containing iron.

For food while the work proceeded, we decided to make four forays [to a nearby village] with every man and horse able to go, and to kill one of our horses every third day to divide among the workers and the sick. Our forays went off as planned. In spite of armed resistance, they netted as much as [100 bushels] of corn.

We had stacks of palmettos gathered, and their husks and fibers twisted and otherwise prepared as a substitute for oakum. A Greek, Don

(continued on next page)

Vocabulary:
concluded = decided
foreseeable = able to be known beforehand
forge = furnace for heating metals
formidable = very difficult
oakum = loose fiber for caulking (filling) seams in wooden ships
pitch = thick, sticky substance used for waterproofing and caulking
rigging = ropes and tackle that control sails
sustain = supply with nourishment

Consider this:
The Spaniards had to use all their skills and imagination to survive. Have you ever had to improvise to solve a serious problem?

Vocabulary:
bellows = device that blasts out air when squeezed, used to make a fire hotter
forays = raids
instigated = brought about, started up
oakum = loose fiber for caulking (filling) seams in wooden ships
palmettos = small palm plants

Commentary:

The Spaniards had landed on the side of Florida bordering the Gulf of Mexico. They called their first Florida campsite "The Cross."

The group had then moved northward, but they were still on the Gulf coast.

The barges (small ships) they built were about 30-32 feet long.

Vocabulary:

ballast = something heavy put deep inside a ship to make it more stable

coves = small sheltered bays in a shoreline

embarked = set out on the journey

flayed = stripped off the skin

junipers = evergreen trees or shrubs

league = 3 miles

pine resins = sticky substance from pine trees

pitch = thick, sticky substance used for waterproofing and caulking

precision = exactness

rigging = ropes and tackle that control sails

Teodoro, made from certain pine resins. Even though we had only one carpenter, work proceeded so rapidly from August 4, when it began, that by September 20 five barges, each 22 elbow-lengths, caulked with palmetto oakum and tarred with pine-pitch, were finished.

From palmetto husks, also horse tails and manes, we braided ropes and rigging; from our shirts we made sails; and from junipers, oars. Such was the country our sins had cast us in that only the most persistent search turned up stones large enough for ballast and anchors. Before this, we had not seen a stone in the whole region. We flayed the horses' legs, tanned the skin, and made leather waterbottles.

Twice in this time, when some of our men went to the coves for shellfish, Indians ambushed them, killing ten men in plain sight of the camp before we could do anything about it. We found their bodies pierced all the way through, although some of them wore good armor. I have already mentioned the power and precision of the Indian archery.

Our pilots estimated, under oath, that from the bay we had named The Cross we had come approximately 280 leagues to this place. In that entire space, by the way, we had seen not a single mountain nor heard of any.

Before we embarked, we lost forty men from disease and hunger, in addition to those killed by Indians. By September 22 all but one of the horses had been consumed. That is the day we embarked …

Source: Cabeza de Vaca, *Adventures in the Unknown Interior of America*. Translated and edited by Cyclone Covey. Albuquerque: University of New Mexico Press, 1998, pp. 45-6.

Setting to Sea in Makeshift Boats

The Spaniards named the place where they built the boats "The Bay of Horses." After making a meal of their last horse, they set out in five roughly built boats, which they called barges. Each barge carried just under fifty men. Cabeza de Vaca was in command of one barge.

Heavily-Loaded Boats

When clothing and supplies were loaded, the sides of the barges remained hardly half a foot above water; and we were jammed in too tight to move. Such is the power of necessity that we should thus hazard a turbulent sea, none of us knowing anything about navigation.

Source: Cabeza de Vaca, *Adventures in the Unknown Interior of America.* Translated and edited by Cyclone Covey. Albuquerque: University of New Mexico Press, 1998, p. 47.

Vocabulary:

hazard = risk

turbulent = moving violently

Swept Away from Land

The Spaniards used materials from some abandoned Indian canoes to raise the sides of their boats a little. As they sailed westward along the coast, water became a huge problem. Unable to find fresh water, some men drank salt water, which made them sick. By this time, they were off the Alabama coast. They all expected to die at any moment.

Suddenly the boats were in fresh water! They had reached the place where a great river—the one we now call the Mississippi—flowed into the sea. Fresh water was all around them. But the strong current created a new problem. It carried their boats far from shore and separated them. The next morning, Cabeza de Vaca could see only two other barges.

The men were rowing now, and exhausted. None of the others could keep up with the captain, since Narváez had kept the best boat and strongest men for himself. They pleaded with the captain to throw them a rope so they wouldn't get separated, but Narváez refused to help. His boat pulled away— and that was the last they ever saw of their captain.

For Cabeza de Vaca, the situation was more desperate than ever.

Consider this:

Compare and contrast Cabeza de Vaca's story with others about ship-wrecks, such as Daniel Defoe's novel *Robinson Crusoe* and the movie *Cast Away*.

Vocabulary:
locomotion = power of movement
parched = roasted

Land at Last

It was winter and bitterly cold, and we had suffered hunger and the heavy beating of the waves for many days. Next day, the men began to collapse. By sunset, all in my barge had fallen over on one another, close to death. Few were any longer conscious. Not five could stand. When night fell, only the navigator and I remained able to tend the barge. ...

Near dawn I seemed to hear breakers resounding... Surprised at this, I called to the navigator, who said he thought we were coming close to land....

As we drifted into shore, a wave caught us and heaved the barge ... out of the water. The jolt when it hit brought the dead-looking men to. Seeing land at hand, they crawled through the surf to some rocks. Here we made a fire and parched some of our corn. We also found rain water. The men began to regain their senses, their locomotion, and their hope.

Source: Cabeza de Vaca, *Adventures in the Unknown Interior of America.* Translated and edited by Cyclone Covey. Albuquerque: University of New Mexico Press, 1998, pp. 45-6.

Help from the Natives in Texas, and Loss of the Barge

Cabeza de Vaca and his men had landed on what is now known as Gavelston island, near the Texas mainland. Fortunately, they fell into the hands of helpful natives who brought them food. The Spaniards soon dug their boat out of the sand and tried to launch it, but the rough waves capsized them. The boat sank, taking several men down with it, and all their supplies and weapons. The Indians took the remaining Spaniards back to their village.

Later, they met up with other Spaniards, survivors of another boat that had capsized nearby. That barge was also beyond repair.

There were now about 80 Spaniards in all on the island. It was November, and getting very cold. Not many of them would survive the winter.

How We Became Medicine-Men

The Islanders wanted to make physicians of us without examination or a review of diplomas. Their method of cure is to blow on the sick, the breath and the laying-on of hands supposedly casting out the infirmity. They insisted we should do this too and be of some use to them. We scoffed at their cures and at the idea we knew how to heal. But they withheld food from us until we complied. An Indian told me I knew not whereof I spoke in saying their methods had no effect. Stones and other things growing about in the fields, he said, had a virtue whereby passing a pebble along the stomach could take away pain and heal; surely extraordinary men like us embodied such powers over nature. Hunger forced us to obey, but disclaiming any responsibility for our failure or success.

An Indian, falling sick, would send for a medicine-man, who would apply his cure. The patient would then give the medicine-man all he had and seek more from his relatives to give. The medicine-man makes incisions over the point of the pain, sucks the wound, and cauterizes it.... I have, as a matter of fact, tried it on myself with good results....

Our method, however, was to bless the sick, breathe upon them ... and pray earnestly to God our Lord for their recovery. When we concluded with the sign of the cross, He willed that our patients should directly spread the news that they had been restored to health.

Source: Cabeza de Vaca, *Adventures in the Unknown Interior of America.* Translated and edited by Cyclone Covey. Albuquerque: University of New Mexico Press, 1998, pp. 64-5.

Commentary:

The Indians suspected that these strange foreigners must have special powers, so they insisted they become healers.

To his surprise, Cabeza de Vaca discovered that some of the native methods worked very well. To his relief, the Spaniards' prayers seemed to work, too.

Vocabulary:

cauterizes = burns to seal
concluded = ended
disclaiming = denying
embodied = included; had
incisions = cuts made into a body or an organ
infirmity = bodily ailment or weakness
physicians = doctors
scoffed = made fun of
virtue = power

From the Island of Doom to Freedom

During that cold and stormy winter, the Spaniards had very little food. By spring, only 15 of the original 80 were still alive. That's why they called the place "Malhado," or the "Island of Doom."

Twelve of those survivors soon left. They crossed to the mainland and headed west, hoping to find their way to other European Christians they knew were in Mexico. But at that time, Cabeza de Vaca was too ill to go with them.

Another Spaniard, Lope de Oviedo, also stayed behind. Oviedo was famous among the Spaniards as the strongest man who had been on Cabeza de Vaca's barge. But he preferred the now-familiar Indian village to the dangers of the unknown. A third man who stayed behind soon died.

Perhaps because of his own illness, Cabeza de Vaca was no longer honored as a healer. When he got well, the Indians made him into a slave. That was a hard life, but Cabeza de Vaca didn't want to abandon Oviedo. Finally, Cabeza de Vaca escaped and became a merchant. Although he traveled far, he always returned to keep in touch with Oviedo.

Commentary:

Cabaza de Vaca was what we would call an entrepreneur—a person who organizes and operates his or her own business

Vocabulary:

barter = swap; trade goods or services without using money

traffic = trade

incessant = unceasing; never ending

ritual = ceremonial; order of doing things in a religious ceremony

sinews = tendons

A Wandering Merchant

I did my best to devise ways of making my traffic profitable so I could get food and good treatment. The various Indians would beg me to go from one quarter to another for things they needed. Their incessant hostilities made it impossible for them to travel cross-country or make any exchanges.

But as a neutral merchant I went into the interior as far as I pleased.

My principal wares were cones and other pieces of sea-snail, conchs used for cutting, sea-beads, and a fruit like a bean which the Indians value very highly, using it for a medicine and for a ritual beverage. This is the sort of thing I carried inland. By barter I got and brought back to the coast skins, red ochre [paint] which they rub on their faces, hard canes for arrows, flint for arrowheads, with sinews and cement to attach them, and tassels of deer hair which they dye red.

This occupation suited me; I could travel where I wished, was not obliged to work, and was not a slave. Wherever I went, the Indians treated me honorably because they liked my commodities.

Source: Cabeza de Vaca, *Adventures in the Unknown Interior of America*. Translated and edited by Cyclone Covey. Albuquerque: University of New Mexico Press, 1998, pp. 66-7.

Vocabulary:
commodities = articles that are traded or sold

News of Others—and a Final Good-bye

Cabeza de Vaca was a merchant for about 22 months in 1528-1532. (Altogether, he was in the Texas coastal area for nearly six years.) While bartering goods, he traveled as far as Oklahoma and back to the coast again. Every year, he visited the island where the Spaniards had been shipwrecked. And each time, he tried to talk Oviedo into leaving.

Finally, in 1532, Oviedo agreed to leave. But the strongman didn't get very far.

Oviedo Finally Leaves—But...

At last, I got him off, across the strait, and across four large streams on the coast; which took some doing, because Oviedo could not swim....

We met some Indians on the other side who ... told us that three men like us lived but a couple of days from here, and said their names. We asked about the others, and were told that they were all dead. Most had died of cold and hunger. But our informants' own tribe had murdered [some of them] for sport....

So we would know they had spoken the truth about the bad treatment of our fellows, they commenced slapping and batting Oviedo and did not spare me either. They would keep throwing clods at us, too, and each of the days we waited there they would stick their arrows

Consider this:
Why do you think the powerful Oviedo was afraid to leave familiar surroundings? After he finally left, why do you think he turned back? Could you write a story from Oviedo's point-of-view?

Vocabulary:
strait = narrow channel that joins two larger bodies of water

(continued on next page)

to our hearts and say they had a mind to kill us the way they had finished our friends. My frightened companion Oviedo said he wanted to go back with the women who had just forded the bay with us (their men having stayed some distance behind). I argued my utmost against such a craven course, but in no way could keep him.

He went back, and I remained alone with those savages.

Source: Cabeza de Vaca, *Adventures in the Unknown Interior of America*. Translated and edited by Cyclone Covey. Albuquerque: University of New Mexico Press, 1998, pp. 68-9.

Castillo, Dorantes, and Estévan

Cabeza de Vaca went with a group of Indians to meet the other Christians. The first Spaniard who came out of their tent was terrified when he saw Cabeza de Vaca. Andrés Dorantes had been on Cabeza de Vaca's barge, but he thought his captain had died on the Island of Doom. The other two had also been on Malhado. Estévan (also called Estéban or Estevánico), a black Moor, had been Dorantes' slave. Alonso de Castillo had arrived on the second barge.

They told Cabeza de Vaca that they were the only men left from the dozen that had left Malhado in 1529. In their travels, Dorantes and the others had seen the remains of two more barges. They later learned that Indians had killed all the men from those boats. They had also come across another Spaniard who said that the Governor, Panfilo de Narváez, and his barge had been carried out to sea. Narváez and the few men with him were never heard from again.

In order to survive, the three Spaniards and the Moor first worked as slaves to the Indians. Then, about September 1534, they escaped. During the following years they walked across what is now Texas, New Mexico, and southeast Colorado.

The explorers did various kinds of work for the Indians they met. They also made things to trade. Like the natives, the four travelers dressed in skins and rags. They suffered from both hunger and thirst. Their diet consisted of prickly pears (the fruit of a kind of cactus) and whatever else they could find or get in trade from the Indians.

Walking Through Texas

We went naked through all this country; not being accustomed to going so, we shed our skins twice a year like snakes. The sun and air raised great painful sores on our chests and shoulders and our heavy loads caused the cords to cut our arms....

I bartered with these Indians in combs I made for them and in bows, arrows, and nets. We made mats, which are what their houses consist of....

Some days the Indians would set me to scraping skins.... I would scrape enough to sustain myself two or three days on the scraps. When [anyone] gave us a piece of meat, we ate it raw.

Source: Cabeza de Vaca, *Adventures in the Unknown Interior of America.* Translated and edited by Cyclone Covey. Albuquerque: University of New Mexico Press, 1998, pp. 92-3.

Consider this:
For these men, survival meant adjusting to many strange new conditions. Can you list some of the most important changes they had to make?

Vocabulary:
bartered = swapped; traded goods without using money

Traveling Medicine Men

The Spaniards and the Moor again became known as healers. Apparently Castillo had a strong religious faith, and he taught the others to use faith healing. Cabeza de Vaca also used other medical skills that he had learned. As their fame spread, native groups welcomed them.

Commentary:

They were now in the general area of what is now Carlsbad, New Mexico. Copper rattles have been found at archeological sites in that region. Each rattle had a pebble inside.

Vocabulary:

foundry = a place with equipment for melting metal and pouring it into molds

league = three miles

magnitude = greatness in size

mica = a mineral found in some rocks, mica splits into flexible sheets

pine nuts = edible seeds of certain pines

prickly pears = the fruit of a kind of cactus

An Operation in New Mexico

We went along the base of the mountains, striking inland more than fifty leagues, at the end of which we came upon forty or so houses.

Among the things the people there gave us was a big copper rattle which they presented Andres Dorantes. It had a face represented on it and the natives prized it highly. They told Dorantes they had received it from their neighbors. Where did *they* get it? It had been brought from the north, where there was a lot of it, replied the natives, who considered copper very valuable. Wherever it came from, we concluded the place must have a foundry to have cast the copper in hollow form.

Departing next morning, we went over a mountain seven leagues in magnitude.

At night we came to many dwellings seated on the banks of a very beautiful stream The residents came halfway out on the trail to greet us, bringing their children on their backs. They gave us many little bags of mica … many beads and cowhide blankets—loading all who attended us with everything they had.

They eat prickly pears and pine nuts; for small pine trees grow in that region with egg-shaped cones whose nuts are better than those of Castile because of their thin husks. … Once our new hosts touched us, they ran back and forth bringing us all kinds of items from their houses for our journey.

They fetched me a man who, they said, had long since been shot in the shoulder through the back and that the arrowhead had lodged above his heart. He said it was very painful and kept him sick. I probed the wound and discovered the arrowhead had passed through the cartilage. With a flint knife I opened the fellow's chest until I could see that the point was sideways and would be difficult to extract. But I cut on and, at last, inserting my knife-point deep, was able to work the arrowhead out with great effort. It was huge. With a deer bone, I further demonstrated my surgical skill with two stitches while blood drenched me, and stanched the flow with hair from a hide. The villagers asked me for the arrowhead, which I gave them. The whole population came to look at it, and they sent it into the back country so the people there could see it.

They celebrated this operation with their customary dances and festivities. Next day, I cut the stitches and the patient was well. My incision appeared only like a crease in the palm of the hand. He said he felt no pain or sensitivity there at all.

Now this cure so inflated our fame all over the region that we could control whatever the inhabitants cherished.

Source: Cabeza de Vaca, *Adventures in the Unknown Interior of America.* Translated and edited by Cyclone Covey. Albuquerque: University of New Mexico Press, 1998, pp. 108-110.

Commentary:

Flint is a very hard type of quartz rock. It can be split into thin, sharp pieces. Pieces of flint were used as tools by many groups of humans before they had metal tools.

Vocabulary:

cartilage = tough connective tissue found in the joints

incision = cut made into a body or an organ

stanched = stopped the flow of blood

Reunited with their Countrymen

In April 1536, Cabeza de Vaca, Castillo, Dorantes, and Estévan made it into central Mexico—eight years after they had first landed in Florida with the Narváez expedition. They were relieved and joyful to find other Spaniards there. But Cabeza de Vaca soon discovered that he had less in common with their countrymen than he had thought.

A Changed Conquistador

Like other Conquistadors, Cabeza de Vaca had once been intent on conquering a New World. But his adventures had changed him. In Mexico, he stopped Spanish slave raids on the Indians. He recommended the use of peaceful means to convert the natives to Christianity—advice that was mostly ignored, especially after he left.

The Spaniards in Mexico tried to explain that they and Cabeza de Vaca's group were all from the same land. The Indians refused to believe them.

Consider this:

The Indians refused to believe that Cabeza de Vaca and his companions were the same species as other Spaniards. This opinion wasn't based on skin color or language. On what did the Indians base their judgements?

Vocabulary:

bestowed = gave;
  presented as a gift
coveted = strongly desired
  something that
  belonged to another

A Different Kind of People

The Indians replied that the Christians lied; We had come from the sunrise, they from the sunset; we healed the sick, they killed the sound; we came naked and barefoot, they clothed, horsed, and lanced; we coveted nothing but gave whatever we were given, while they robbed whomever they found and bestowed nothing on anyone.... To the last I could not convince the Indians that we were of the same people as the Christian slavers.

Source: Cabeza de Vaca, *Adventures in the Unknown Interior of America.* Translated and edited by Cyclone Covey. Albuquerque: University of New Mexico Press, 1998, p. 128.

After the Journey

Cabeza de Vaca returned to Spain in 1537. Castillo also went to Spain, but then returned to Mexico City, where he married and became a citizen. Dorantes had hoped to go on another expedition with Cabeza de Vaca. When that didn't work out, he married a rich widow and settled down. Estévan joined another group of explorers, as their guide. (See page 39 for the story of that journey.)

Later, Cabeza de Vaca was put in charge of an expedition to South America, where he continued to prohibit harsh treatment of the Indians. For this he was arrested by other Spaniards and returned to Spain in chains. His jail sentence was eventually annulled, and he held a position of honor until his death in 1557.

Hernando de Soto

Hernando de Soto

Hernando de Soto (about 1500-1542) grew up in Spain, listening to stories from explorers who had been to the New World. He was eager to follow their example. As a young teenager, he sailed to Central America with a newly appointed Spanish administrator. After expeditions in Nicaragua and Peru, de Soto returned to Spain, where he became a favorite of the court.

In 1537, de Soto was named governor of Cuba and given official permission to conquer Florida. Just after that, another explorer with the same ambitions returned to the Spanish court.

Hearing from Cabeza de Vaca

After Don Hernando had obtained the concession, a fidalgo arrived at Court from the Indias, Cabeça de Vaca by name, who had been in Florida with Narváez; and he stated how he with four others had escaped, taking the way to New Spain; that the Governor had been lost in the sea, and the rest were all dead.

He brought with him a written relation of adventures,....

Source: a transcription by Dr. Jon Muller, SIU, of *The Account of Elvas,* translated by Buckingham Smith. New York: Allerton Book Co., Mcmxxii. Found at http://www.floridahistory.com/elvas1.html

Commentary:
Several men who traveled with de Soto kept journals. One was simply referred to as "an unnamed Portuguese Officer," or "The Gentleman of Elvas." Parts of his journal are included in this section.

Vocabulary:
Cabeça = alternate spelling of Cabeza
concession = land granted for business use
fidalgo = nobleman

De Soto and Cabeza de Vaca

Cabeza de Vaca told de Soto about the New World. He described the poverty he had seen among the natives. However, when pressured, de Vaca also gave the impression that wealth could be found there. De Soto wanted Cabeza de Vaca to go back with him, as second in command of the expedition. But after his experiences on the last voyage, Cabeza de Vaca wasn't interested in being second in command to anybody.

De Soto had plenty of other volunteers for the voyage. They arrived in Cuba in June, 1538. De Soto spent about a year getting things organized for the next stage of the journey. He built up a supply of food and other necessities—including chains and collars for human beings. Since he had talked to Cabeza de Vaca, de Soto avoided some mistakes Narváez had made on that earlier expedition. For example, in addition to conquistadors and mercenary footsoldiers, de Soto took along craftsmen who could build whatever might be needed.

He finally set sail for Florida with about 650 men, an additional hundred or so camp followers, and many slaves. With them, they took horses, hogs, and attack dogs.

Vocabulary:

Adelantado = captain-
  general; leader
league = 3 miles
shoals = shallow places
  in a body of water
caravels = type of ship
pinnaces = type of ships
nigh = near

De Soto Sails to Florida

On Sunday, the 18th day of May, in the year 1539, the Adelantado sailed from Havana with a fleet of nine vessels, five ... ships, two caravels, two pinnaces; and he ran seven days with favourable weather. On the 25th of the month ... the land was seen, and anchor cast a league from shore, because of the shoals. On Friday, the 30th, the army landed in Florida, two leagues from the town of an Indian chief named Utica. Two hundred and thirteen horses were set on shore, to unburden the ships, that they should draw the less water; the seamen only remained on board....

So soon as the people were come to land, the camp was pitched on the sea-side, nigh the bay, which goes up close to the town. Presently the Captain-General ... taking seven horsemen with him, beat up the country half a league

Florida tribesmen, of the Timucua tribe, on a military expedition.

about, and discovered six Indians, who tried to resist him with arrows, the weapons they are accustomed to use. The horsemen killed two, and the four others escaped, the country being obstructed by bushes and ponds, in which the horses bogged and fell, with their riders, of weakness from the voyage. At night the Governor, with a hundred men in the pinnaces, came upon a deserted town; for, so soon as the Christians appeared in sight of land, they were descried, and all along on the coast many smokes were seen to rise, which the Indians make to warn one another....

Source: a transcription by Dr. Jon Muller, SIU, of *The Account of Elvas,* translated by Buckingham Smith. New York: Allerton Book Co., Mcmxxii. Found at http://www.floridahistory.com/elvas1.html

Consider this:
To communicate with their ships or with other groups of explorers, the Spaniards could only send human messengers. The Indians had a far better system—smoke signals.

Vocabulary:
bogged = bogged down; sinking in mud
descried = spotted; discovered
obstructed = blocked
pinnaces = type of ship

De Soto and the Natives

The Indians had good reason to desert their homes when they knew Spaniards were coming. Conquistadors often plundered towns, taking food and women. In order to get whatever they wanted, the Spaniards would usually capture the chief and hold him for ransom. They forced captured Indians to serve as guides and translators. When de Soto's guides managed to escape, the Spaniards went looking for more. However, they often met with fierce resistance from native warriors.

Commentary:

Even though the Spaniards had guns and wore armor, they discovered that Indian weapons and fighting skills were extremely effective against them.

Vocabulary:

arquebuse = gun that fired a large round lead ball

nimble = quick; agile

traversing = crossing; passing

Indian Warriors

The Indians are exceedingly ready with their weapons, and so warlike and nimble, that they have no fear of footmen; for if these charge them they flee, and when they turn their backs they are presently upon them. They avoid nothing more easily than the flight of an arrow. They never remain quiet, but are continually running, traversing from place to place, so that neither crossbow nor arquebuse can be aimed at them. Before a Christian can make a single shot with either, an Indian will discharge three or four arrows; and he seldom misses of his object. Where the arrow meets with no armour, it pierces as deeply as the shaft from a crossbow. Their bows are very perfect; the arrows are made of certain canes, like reeds, very heavy, and so stiff that one of them, when sharpened, will pass through a target. Some are pointed with the bone of a fish, sharp and like a chisel; others with some stone like a point of diamond: of such the greater number, when they strike upon armour, break at the place the parts are put together; those of cane split, and will enter a shirt of mail, doing more injury than when armed.

Source: a transcription by Dr. Jon Muller, SIU, of *The Account of Elvas*, translated by Buckingham Smith. New York: Allerton Book Co., Mcmxxii. Found at http://www.floridahistory.com/elvas1.html

Another Survivor of the Narváez Expedition

Early in his travels, de Soto came across a European living among the Indians. Like Cabeza de Vaca, Juan Ortiz had been on the ill-fated Narváez expedition. Ortiz had made it back to Cuba, then returned with new men and a new ship to see if he could find any survivors.

Ortiz and his men were taken captive by the Indian chief Utica. According to Elvas, the chief's daughter talked her father out of killing Ortiz: "Though one Christian, she said, might do no good, certainly he could do no harm, and it would be an honour to have one for a captive...." (Does that story sound familiar? Some historians believe that John Smith later borrowed the Ortiz story to liven up his own tales of Indians in Virginia. The better-known version wasn't even written until after Pocahontas had died.)

Ortiz eventually escaped Utica and lived with another tribe. When the Spaniards found him, he had been among the Indians for 12 years.

Finding Juan Ortiz

When Baltasar de Gallegos came into the open field, he discovered ten or eleven Indians, among whom was a Christian, naked and sunburnt, his arms tattooed after their manner, and he in no respect differing from them. As soon as the horsemen came in sight, they ran upon the Indians, who fled, hiding themselves in a thicket, though not before two or three of them were overtaken and wounded. The Christian, seeing a horseman coming upon him with a lance, began to cry out: " Do not kill me, cavalier; I am a Christian! Do not slay these people; they have given me my life!"

Source: a transcription by Dr. Jon Muller, SIU, of *The Account of Elvas,* translated by Buckingham Smith. New York: Allerton Book Co., Mcmxxii. Found at http://www.floridahistory.com/elvas1.html

Consider this:
In what ways does the Ortiz story resemble Cabeza de Vaca's experiences?

Vocabulary:
cavalier = mounted soldier; knight

De Soto and his troops traveled through what are now the states of Florida, Georgia, the Carolinas, Tennessee, Alabama, Mississippi, and Arkansas. (U.S. Department of the Interior, De Soto National Memorial)

The End of the Expedition

De Soto and his army spent their first winter in Florida. De Soto was paying for this expedition himself. In return, he was to keep whatever wealth he discovered, minus 5% for the Spanish crown. But he gained no real wealth from the entire journey.

In the spring of 1540, de Soto and his army went north into what is now Georgia and the Carolinas. Then they headed southwest. Indian chiefs kept the Spaniards moving by telling stories about gold just ahead in the next village. The Spaniards fought their way to Texas and back, burning Indian villages and losing their own men to battle and disease.

In May 1542, de Soto's expedition returned to Mississippi, where de Soto died of a fever (probably malaria). The remaining 322 members of the expedition built boats and launched them on the Mississippi River. They finally made it to the Spanish town of Vera Cruz in Mexico.

So de Soto died in the New World, and the survivors of the expedition returned without any riches. The expedition was judged to be a failure. But the information they brought back about the southeast would become more valuable as time went on.

Francisco Vásquez de Coronado

Francisco Vásquez de Coronado

Born into the Spanish nobility, Coronado (1510-1554) traveled to the New World when he was 25. In Mexico City, he married a wealthy woman, was put on the city council, and was made governor of a province. But Coronado wasn't satisfied. He was fascinated by stories he heard about wealth to be found in the north.

Coronado's story was told in a journal written by a man who was on the expedition. The journal-keeper waited many years to publish the story, but finally decided it was important to set the record straight.

Conflicting Stories

I think that the twenty years and more since that expedition took place have been the cause of some stories which are related. For example, some make it an uninhabitable country, others have it bordering on Florida, and still others on Greater India…. They are unable to give any basis upon which to found their statements. There are those who tell about some very peculiar animals, who are contradicted by others who were on the expedition, declaring that there was nothing of the sort seen. Others differ as to the limits of the provinces and even in regard to

(continued on next page)

Commentary:

Twenty years after Coronado's expedition, the New World was still unknown territory to most people. It was still the subject of myths and legends.

Vocabulary:

provinces = territory governed by another country

the ceremonies and customs, attributing what pertains to one people to others. All this has had a large part, my very noble lord, in making me wish to give now, although somewhat late, a short general account for all those who pride themselves on this noble curiosity, and to save myself the time taken up by these solicitations.

Source: Pedro de Castaneda, *Account of the Expedition to Cíbola*. Found at http://www.pbs.org/weta/thewest/resources/archives/one/corona1.htm

Meeting Cabeza de Vaca in Mexico

It happened that just at this time three Spaniards, named Cabeza de Vaca, Dorantes, and Castillo Maldonado, and a negro, who had been lost on the expedition which Panfilo de Narváez led into Florida, reached Mexico. They came out through Culiacan, having crossed the country from sea to sea, as anyone who wishes may find out for himself by an account which this same Cabeza de Vaca wrote and dedicated to Prince Don Philip, who is now King of Spain and our sovereign. They gave the good Don Antonio de Mendoza an extended account of some powerful villages, four and five stories high, of which they had heard a great deal in the countries they had crossed, and other things very different from what turned out to be the truth.

Source: Pedro de Castaneda, *Account of the Expedition to Cíbola*. Found at http://www.pbs.org/weta/thewest/resources/archives/one/corona1.htm

Searching for Cíbola

The Spaniards had heard a legend of seven fabulous cities. According to that ancient story, seven Catholic bishops had crossed the Atlantic Ocean in the year 714. They had each founded a city, and the seven cities had grown grand and wealthy.

Cabeza de Vaca had also heard those stories. He asked many Indians about the cities, and he thought they might actually exist. In 1539—about the same time that de Soto was establishing himself in Florida—Coronado set out to find the the Seven Cities of Cíbola.

Coronado made his headquarters in northern Mexico. The first group to push ahead toward Cíbola was led by Fray Marcos, a French friar who spoke a little Spanish but no native languages. Their scout was an explorer who had been to that region before.

The Death of Estévan

It seems that, after the friars ... and the negro had started, the negro did not get on well with the friars.... Besides, the Indians in those places through which they went got along with the negro better, because they had seen him before. This was the reason he was sent on ahead to open up the way and pacify the Indians....

But [in Cíbola] ... the older men and the governors heard his story and took steps to find out the reason he had come to that country.... The account which the negro gave them of two white men who were following him, sent by a great lord, who knew about the things in the sky, and how these were coming to instruct them in divine matters, made them think that he must be a spy or a guide from some nations who wished to come and conquer them.... Besides these other reasons, they thought it was hard of him to ask them for turquoises and women, and so they decided to kill him.

Source: Pedro de Castaneda, *Account of the Expedition to Cíbola*. Found at http://www.pbs.org/weta/thewest/ resources/archives/one/corona1.htm

Commentary:

According to some accounts, Estévan went on ahead because he was unwilling to wait for Fray Marcos. The Moor knew the territory well, and he no longer considered himself a slave.

Fray Marcos' Story

Going ahead of the main group, Estévan had reached the Zuñi Indian pueblos in what is now western New Mexico. The town was a major trading center, where goods such as buffalo hides, cotton, turquoise, coral, carved seashells, and Macaw parrots or their feathers could be found. It was not, however, the fabulously wealthy Cíbola of ancient legend.

Indian stories about the grand cities aren't hard to explain. The pueblos were, in fact, much more impressive than the smaller villages many Indians lived in. But it's harder to explain the report that Fray Marcos took back. The priest told Coronado that he had actually seen the golden cities, and that the smallest of them was larger than Mexico City.

Some say that Fray Marcos saw the pueblos from a distance and mistook them for grander cities. Others suggest that he simply turned back when he heard that Estévan had been killed, and never saw the pueblos at all—which he didn't want to admit to Coronado.

In 1540—as de Soto was searching for gold in Georgia and Alabama— Coronado reached the pueblos. The Spaniards soon realized that they had not discovered the Seven Cities of Cíbola.

Commentary:

Fray Marcos returned with this expedition. There is no record of how he explained the differences in what he reported and what they actually found.

The Cities the Spaniards Found

The next day they entered the settled country in good order, and when they saw the first village, which was Cíbola, such were the curses that some hurled at Fray Marcos that I pray to God He may protect him from them.

It is a little, crowded village, looking as if it had been crumpled all up together. There are ranch houses in New Spain which make a better appearance at a distance. It is a village of about 200 warriors, is three and four stories high, with the houses small and having only a few rooms, and without a courtyard. One yard serves for each section. The people of the whole district had collected here, for there are seven villages in the province, and some of the others are even larger and stronger than Cíbola. These folk waited for the army, drawn up by divisions in front of the village. When they refused to have peace on the terms the interpreters extended to them, but appeared defiant … they were at once put to flight. The Spaniards then attacked the village, which was taken with not a little difficulty, since they held the narrow and crooked entrance. During the attack they knocked the general down with a large stone, and would have killed him but for [two soldiers] who threw themselves above him and drew him away, receiving the blows of the stones, which were not few. But the first fury of the Spaniards could not be resisted, and in less than an hour they entered the village and captured it. They discovered food there, which was the thing they were most in need of. After this the whole province was at peace.

Source: Pedro de Castaneda, *Account of the Expedition to Cíbola.* Found at http://www.pbs.org/weta/thewest/resources/archives/one/corona1.htm

Commentary:
Coronado was wearing gilded armor, making it hard for him to run. It also made him easy to see—perhaps why he was so quickly hit with a large stone.

Vocabulary:
province = territory governed by another country

No Wealth to be Found

Fray Marcos wasn't very popular with the men of the expedition. He was also very tired, so he accompanied a group of messengers back to Mexico.

Coronado continued his search for Cíbola. He reached what is now Kansas, and sent out smaller parties that explored other areas. These groups saw much of New Mexico and went as far west as the Arizona/California border. In 1542, Coronado returned to Mexico, and his expedition was declared a failure. Like de Soto, Coronado would only later be recognized for the value of his explorations.

Some of the men that Coronado sent out on shorter scouting trips also made important discoveries.

Commentary:

Coronado thought that he and his men would be traveling near the coast, but he was mistaken. Like Narváez, Coronado was never able to reconnect with his ship. Meanwhile, Alarcón made valuable notes about Indians he met. He also discovered that Baja California wasn't an island—as the Spaniards had thought—but a peninsula.

The Voyage of Don Pedro de Alarcón

After the whole force had left Mexico, [Coronado] ordered Don Pedro de Alarcón to set sail with two ships … to take the baggage which the soldiers were unable to carry, and … to sail along the coast near the army, because he had understood from the reports that they would have to go through the country near the seacoast, & that we could find the harbors by means of the rivers, and that the ships could always get news of the army, which turned out afterward to be false.…

Source: Pedro de Castaneda, *Account of the Expedition to Cíbola*. Found at http://www.pbs.org/weta/thewest/resources/archives/one/corona1.htm

Alarcón sailed from the west coast of Mexico up through the Gulf of California and into the Colorado River. Unable to locate Coronado, Alarcón finally gave up. But he buried a message at the foot of a tree on the Gulf coast.

One of Coronado's men, Melchior Díaz, later explored that region. Indians told him that another Spaniard had left some letters. They showed Díaz where a message was carved on a tree: "Alarcón reached this place; there are letters at the foot of this tree." The buried letters explained that Alarcón had waited for news, but had finally returned to Mexico.

On the Edge of the Grand Canyon

In 1540, Captain Garcia Lopez de Cardenas, became the first European to see the Grand Canyon. Sent northward by Coronado to search for Cíbola, Cardenas and his party spent three days on the Canyon rim. They tried to get down to the river at the bottom of the gorge, but ran short of supplies and had to give up.

Larger Than It Looks

They spent three days on this bank looking for a passage down to the river, which looked from above as if the water was six feet across, although the Indians said it was half a league wide…. [After three days] the three lightest and most agile men, made an attempt to go down at the least difficult place, and went down until those who were above were unable to keep sight of them. They returned about four o'clock in the afternoon, not having succeeded in reaching the bottom on account of the great difficulties which they found, because what seemed to be easy from above was not so, but instead very hard and difficult. They said that they had been down about a third of the way and that the river seemed very large from the place which they reached…. Those who stayed above had estimated that some huge rocks on the sides of the cliffs seemed to be about as tall as a man, but those who went down swore that when they reached these rocks they were bigger than the great tower of Seville.

Source: Pedro de Castaneda, *Account of the Expedition to Cíbola.* Found at http://www.pbs.org/weta/thewest/resources/archives/one/corona1.htm

Consider this:

Why do you think the men had such a hard time judging the size of things they were looking at?

Vocabulary:

agile = quick, nimble
estimated = judged; calculated
league = three miles

Commentary:

Seville is a city in Spain that has many ancient Moorish buildings. The tower referred to is probably the "Giralda," a minaret (a tower on a mosque) built in the 12th century. It is more than 300 feet tall.

Settlers and Missionaries

Spanish explorations of the New World continued by land and sea. However, Spanish authorities soon realized that no rich empire existed in what is now the United States. They weren't going to find a source of real wealth north of the Rio Grande. However, not all Spaniards came to the New World in search of gold. Some conquistadors were accompanied by men, women, and even children—people who intended to stay.

Spain began to view their northern territory as a protective barrier for their settlements in Mexico. They also saw it as a good place for saving pagan souls. The conquistadors were always accompanied by one or more priests or friars—missionaries who came to convert the natives to Christianity.

In Florida, the priests began building missions—places where they could carry on their work—soon after St. Augustine was founded. Catholic missions were also founded in Texas, New Mexico, Arizona, and California. To defend the missions and the border, Spanish troops were stationed in fortified garrisons called presidios.

In the 1770's, Fray Junipero Serra was a powerful voice arguing for the establishment of missions in the New World. He especially influenced the building of missions in California.

Spanish mission

Juan Rodríguez Cabrillo was the first to explore the California coast-line. As a young man, he arrived in Cuba in the 1500s. Cabrillo joined the Narváez expedition, then switched to Cortés' troops in Mexico. In 1542, he was in command of three ships that sailed northward along the coast of Baja California. They made it north about as far as where San Francisco is now, where it is believed that Cabrillo died in 1543. (From Cabrillo's Log, 1542-1543. Found at www.mms.gov/omm/pacific)

Fray Sierra–Missions Before Towns

Missions, my lord, missions-that is what this country needs. They will not only provide it with what is most important—the light of the Holy, Gospel—but also will be the means of supplying foodstuffs for themselves and the Royal Presidios. They will accomplish this far more efficiently than these pueblos without priests....

Later on, when the [all] have become Christian.... I assure you that then will be the proper time for introducing towns of Spaniards. Let them be of good conduct and blameless life.

Source: Winefred E., Wise, *Fray Junipero Serra and the California Conquest,* Charles Scribner's Sons, NY: 1967, p. 150.

Consider this:

Why do you think that Fray (Friar) Sierra was eager to have religious centers established before regular towns?

Vocabulary:

Presidios = military forts

Revolt of the Pueblos in New Mexico

By August 1680, there were 2800 Spanish settlers in New Mexico. The military leader and governor at Santa Fe, Don Antonio de Otermín, began to receive reports of the deaths of priests and settlers. The Indians were angry over arrests of their medicine men and the suppression of native religions. Santa Fe was soon under siege, and Otermín recognized one of the enemy.

Summary:

He came to see me, and I asked if he had gone crazy. He spoke our language. He had lived among Spaniards all his life. I had put my confidence in him. Now he came to us as a rebel leader. He replied that they had elected him.

They were carrying a white flag and a red one. The white was for peace, and the red for war. If we wanted the white (and peace), we must agree to leave the country. If we wanted the red, we would die.

The Pueblo Revolt

He came to where I was, and ... I asked him how it was that he had gone crazy too—being an Indian who spoke our language, was so intelligent, and had lived all his life in the villa among the Spaniards, where I had placed such confidence in him—and was now coming as a leader of the Indian rebels. He replied to me that they had elected him as their captain, and that they were carrying two banners, one white and the other red, and that the white one signified peace and the red one war. Thus if we wished to choose the white it must be upon our agreeing to leave the country, and if we chose the red, we must perish....

Source: Translation from C. W. Hackett, ed., Historical Documents relating to New Mexico, Nueva Vizcaya, and Approaches Thereto, to 1773, vI. . III [Washington: Carnegie Institution of Washington,, 937] pp. 327-35. Found at http://www.pbs.org/weta/thewest/resources/archives/one/pueblo.htm

Otermín chose the red, and many of his soldiers were killed or wounded during the battles that followed. There were also many women and children in Santa Fe, and supplies were running low. Otermín and his troops fought their way out, but the army and the settlers were driven from one town to the next. Every settlement was under attack by well-organized Pueblo warriors.

The Spanish survivors gathered near El Paso. There were just 1946 people left, and only 155 were armed soldiers. They retreated across the Rio Grande into Mexico.

In the 1690s, the Spanish regained control of New Mexico.

The Missions of Texas

Spanish settlement in Texas began slowly in the 1650s, with just a few priests and temporary churches. In the 1680s, the French also established forts in that area. By the early 1700s, the Spanish had decided to get rid of any lingering Frenchmen and settle Texas themselves.

In 1690, Father Damian Massanet accompanied the explorer Captain Alonso de Leon into east Texas. They were to destroy the remains of a French fort, and to found a permanent mission in that region.

Founding a Mission

We left [Mexico] for the Tejas on the third day of the Easter feast, March 28, '90. When we left, the [expected] twenty soldiers ... had not yet arrived. The forty [people with us] were for the most part tailors, shoemakers, masons, miners—in short, none of them could catch the horses on which they were to ride that day, for when they had once let them go they could manage them no longer. Besides, we had saddles that could not have been worse....

[They arrived at their destination in May.] The next morning, I went out with Captain Alonso de Leon a little way, and found a delightful spot close to the brook, fine woods, with plum trees like those in Spain. And soon afterwards, on the same day, they began to fell trees and cart the wood, and within three days we had a roomy dwelling and a church wherein to say mass with all propriety. We set in front of the church a very high cross of carved wood.

Source: Lilia M. Casis, translator, "Letter of Don Damian Manzanet to Don Carlos de Siguenza Relative to the Discovery of the Bay of Espiritu Santo," *The Quarterly of the Texas State Historical Association,* 11(1899), 293-309. Found at http://www.texasbob.com/texdoc2.html

Vocabulary:

with all propriety = appropriately; suitably

Tejas = Texas

Mission Life in 1786 California

Life in the missions was highly regulated for the Christian Indians who lived there. The work was hard, and the meals might seem a bit skimpy to many of us today. For more information, see *The Missions of California,* by Phyllis Raybin Emert (Discovery Enterprises, 1997).

Commentary:

Although the food in the missions was simple, for some Indians it might have been an improvement over what they had in their native villages. Explorers such as Cabeza de Vaca often commented on the starvation conditions of the natives in some areas.

Vocabulary:

insipid = tasteless
domestic = related to the house or household

A Day in the Mission

[All] rise with the sun, and immediately go to prayers and mass…. [Breakfast is] a kind of soup, made of barley meal…. called *atole*. They eat it without either butter or salt, and it would certainly to us be a most insipid mess.

…after which they all go to work, some to till the ground with oxen, some to dig in the garden, while others are employed in domestic occupations, all under the eye of one or two missionaries….

At noon the bells give notice of the time of dinner…. This second soup is thicker than the former, and contains a mixture of wheat, maize, peas, and beans; the Indians call it po-zo1e…. Evening prayer … is followed by a distribution of *atole,* the same as at breakfast….

On high festivals an allowance of beef is distributed which many eat raw, particularly the fat, considered by them as delicious as the finest butter or the most excellent cheese….

… The holy fathers have constituted themselves guardians of the virtue of the women. An hour after supper, they take care to secure all the women whose husbands are absent, as well as the young girls above the age of nine years, by locking them up, and during the day they entrust them to the care of elderly women….

Source: Malcolm Margolin, Life in a California Mission—*The Journals of Jean Francois de la Pérouse Monterey,* Berkeley, CA, 1786, pp. 85-91, 95-6.

Fort Mose—Sanctuary for Slaves

One of the most interesting of the Spanish settlements was a sanctuary and fortress. As the English and Spanish struggled for control of the deep south, Spain offered religious sanctuary to slaves who ran away from English plantations. By 1738, 100 Africans had made their way to St. Augustine—often with the help of Indians. The Spaniards then established a fort and a black militia to defend from possible English invasion. More Africans came, forming families, and becoming farmers and craftspeople. Ft. Mose was abandoned in 1763, when Florida became an English colony.

The Fort in 1759

The Fort at Mose is situated on the banks of the River which runs to the north ... the part that faces the river has no protection of defense whatsoever and is formed by two small bastions which look landward on which are mounted two four-pound cannons and six swivel guns divided among them ... the earthwork embankment is covered with thorns and the moat is three feet wide and two feet deep ... the housing ... includes ... some huts of thatch ... and the sacristy ... in which the priest lives, is a very small room and serves as the chapel for the fort.

Source: Report of Father Juan Joseph de Solana, from Kathleen Deagan and Darcie MacMahon, *Fort Mose: Colonial America's Black Fortress of Freedom,* University Press of Florida, 1995.

Commentary:

One reason the Spaniards gave sanctuary to runaways was to convert them to Catholicism. It was also a way of striking at their English enemies.

Spaniards also held many slaves, but their slavery laws were different from those in other countries. For example, Spanish slaves could earn money and buy their freedom; it was illegal to break up slave families; and slaves could sue their owners in court.

Afterword: Our Spanish Heritage

Native Americans puzzled over Spanish explorers. Spaniards and Indians often fought each other, but sometimes they learned from each other. Images such as this pictograph, found at Canyon de Chelly, Arizona, show that Indians had observed conquistadors in Arizona. (Found at PBS.org/weta/thewest/resources/archives)

During the 16th century, Spain was the most powerful country in Europe. Her ships ruled the Atlantic Ocean, bringing back incredible riches from the conquered Aztec and Inca nations. In 1588, the Spanish Armada (fleet of warships) was considered invincible. Philip II, the king of Spain planned to use his Armada to invade England.

However, England had also built a powerful navy. And the Dutch, who were struggling to free themselves from Spanish rule, had their own highly maneuverable ships. When the Spanish Armada tried to attack, it was defeated by a combination of English and Dutch naval forces. Although they built a new fleet, Spanish power began to decline. Dutch, French, and English privateers began to raid gold-laden Spanish ships on their way back from the New World. (A privateer was a privately owned ship that was authorized by a government to attack and capture enemy vessels.)

By the next century, power had shifted to England, and England began to colonize North America. We generally trace our United States history to those early English colonies. But we also have a Spanish heritage, especially in the American southwest and on the west coast. We can still find signs of those earlier explorers who went so boldly into completely unknown territory.

Research Activities/Things to Do

- Study current maps of Florida, Georgia, Alabama, Mississippi, Louisiana, Texas, New Mexico, Arizona, or California to find places that were named by or after the Spanish explorers of the 16th century.

- In New England and the Mid-Atlantic states, many of the geographic names are British, French, German, and Dutch. Why do you think that is?

- When the Spanish and other European explorers came to North and South America looking for riches and assumed that they would convert the Native Americans to Christianity, how were they imposing their values on the people whom they found living here? Make a list of the many ways they planned to change life as they found it here. What was successful? What didn't work out for the Conquistadores? Explain your answer.

- Choose one of the Spanish explorers who came to the "New World" in the 16th century and compare him to an astronaut. Include answers to the following questions, as well as your own comparisons.

  - Who planned the trip?

  - Who paid for the trip?

  - How many others went along?

  - What was the purpose of the expedition?

  - What were the difficulties encountered?

  - Was the expedition repeated at a later time?

  - Name three things that were learned as a result of the exploration.

  - List the long-range and immediate benefits of the trip.

  - List the problems that arose.

  - Describe food, clothing, transportation of the participants.

  - How easily did the explorers return home, if they did return home.

  - Whose trip was easier for the people involved, and why.

- Following is an excerpt from a journal of one member of deSoto's party, as the men traveled through Texas. Using the form on the opposite page, analyze the document below.

Austin was the end of the road for deSoto's army. Scouting parties were sent out, in several directions, to explore under Harvest Moon: one west, up the Colorado River through the Texas Hill Country, the other southwest, to San Antonio.

Final Expeditions

"There [at Austin] the Indians told them that 10 days' journey thence toward the west was a river called Daycao, where they sometimes went to hunt in the mountains and kill deer [probably near the Llano/Colorado River junction]; and that on the other side [of the mountains] they had seen people, but did not know what village it was [probably Llano Indians]. There [at Austin] the Christians took what corn they found and could carry [on the scouting parties] and after marching for ten days through an unpeopled region reached the [Llano] river of which the Indians had spoken." They found a poor village and brought back two captives "to the river where the governor was awaiting them [on the Colorado River at Austin]. They continued to question the Indians in order to learn from them the population." to the westward, but there was no Indian in the camp who understood their [Llano] language."

. . . .

"On this last journey that our people made after the death of Governor Hernando de Soto they traveled, going and returning, and counting the expedition that the scouts made [beyond Austin], more than 350 leagues [900+ miles, a remarkably accurate measure], during which a hundred Spaniards and eighty horses died at the hands of the enemy and from sickness..."

Source: http://floridahistory.com/texas.html#texas

Written Document Worksheet

1. **Type of document:**

   ❏ Newspaper   ❏ Diary/Journal   ❏ Letter

   ❏ Ad   ❏ Telegram   ❏ Patent

   ❏ Deed   ❏ Census Report   ❏ Memo

   ❏ Report   ❏ Other _____

2. **Clues about the Document:**

   ❏ Interesting Stationery   ❏ "RECEIVED"

   ❏ Fold Marks   ❏ "CLASSIFIED"

   ❏ Handwritten   ❏ Written Notations

   ❏ Typed   ❏ "TOP SECRET" Stamp

   ❏ "Copy"   ❏ Official

3. **Date(s) of Document:**   ❏ No Date

4. **Author of Document:**   **Position:**

5. **For what audience was the document written?**

6. **Key information** *(What do you think are the three most important points?)*

   1.

   2.

   3.

7. **Choose a quote from the document that helped you to know why it was written.**

8. **Write down two clues which you got from the document that tell you something about life in the U.S. at the time it was written.**

   1.

   2.

9. **Write a question to the author that you feel is unanswered in the document.**

- Using the form on the facing page, compare the two 16th century engravings of Natives in the "New World" by Theodore De Bry shown here.

- What are some of the Natives doing?

- Describe the farming and other activities on land.

Analyzing Graphics Worksheet

Some or all of the following will help you to understand graphics. Use the worksheet to jot down notes about the pictures you are looking at. You may find this page helpful when looking at other graphics, too.

1. **What is the subject matter?**

2. **What details provide clues?** *(Check each box that applies to the picture you are studying)*

    ❏ Scene   ❏ Buildings   ❏ People

    ❏ Clothing   ❏ Artifacts   ❏ Time of day

    ❏ Written Message   ❏ Season   ❏ Activity taking place

    ❏ Other _____

3. **Can you tell what the date is? If there is no date, can you guess when it was probably illustrated?**

4. **What is the purpose of the art?**

    ❏ Private use   ❏ Recording an event   ❏ News story

    ❏ Art   ❏ Advertising   ❏ Other _____

5. **Can you tell anything about the point of view of the graphic?** *(Is there a message for the person looking at the illustration?)*

6. **What details make this illustration interesting?**

7. **What can you learn about the people who lived at this time or in this place?**

8. **List the things in the pictures that you have done, and then list the things that you have never done.**

Suggested Further Reading

Deitch, Kenneth M. *A Guide to Teaching about the Columbus Controversy*, Carlisle, MA: Discovery Enterprises, Ltd., 1991.

Faber, Harold. *The Discoverers of America*, Scribner, 1992.

Morris, John M. *From Coronado to Escalante: The Explorers of the Spanish Southwest*, Chelsea House, 1992.

Fernández, José, and Martin, Favata. *The Account: Alvar Núñez Cabeza de Vaca's Relacion,* Arte Público Press, 1993 (New translation).

Formisano, Luciano, ed. *Letters from a New World: Amerigo Vespucci's Discovery of America*, translated by David Jacobson, New York: Marsilio, 1992.

Koning, Hans. *The Conquest of America: How the Indian Nations Lost Their Continent*, New York: Monthly Press, 1993.

Stanush, Barbara Evans. *Texans: A Story of Texan Cultures for Young People*, University of Texas Institute of Texan Cultures, 1998.

Wade, Mary Dodson. *Cabeza de Vaca: Conquistador Who Cared*, Colophon House, 1994.

—— *Esteban: Walking Across America*, Colophon House, 1995.

Weisman, Joanne B., and Deitch, Kenneth M. *Christopher Columbus and the Great Voyage of Discovery,* Carlisle, MA: Discovery Enterprises, Ltd., 1990.

Wills, Charles. *A Historical Album of Texas*, Millbrook Press, 1995.

Suggested Web Sites

http://groups.msn.com/SpainInAmerica/indios
http://www.geocities.com
http://www.enchantedlearning.com/explorers
http://www.mariner.org/age/index

Spanish Missions:
http://flspmissions.tripod.com/index.ht
http://employees.oneonta.edu/jacksorh/spanishmissions
http://www.californiamissions.com